THE ASTRONAUT'S WINDOW

THE ASTRONAUT'S WINDOW

Collection of Poems and Short
Stories Celebrating Nature

HAZEL ANN LEE

THE ASTRONAUT'S WINDOW
COLLECTION OF POEMS AND SHORT STORIES CELEBRATING NATURE

iUniverse books may be ordered through booksellers or by contacting:

iUniverse
1663 Liberty Drive
Bloomington, IN 47403
www.iuniverse.com
1-800-Authors (1-800-288-4677)

ISBN: 978-1-5320-9565-8 (sc)
ISBN: 978-1-5320-9566-5 (e)

Library of Congress Control Number: 2020903427

First Edition

Print information available on the last page.

iUniverse rev. date: 02/26/2020

Contents

To my family especially my mother, Mrs. Grace Garner Lee, and my sister, Dr. Cynthia Cozette Lee, my friends, the scientists, teachers and mentors who inspired me and potential scientists dedicated to scientifically explaining and understanding the vast workings of the universe.

Preface

When I was ten years old, I watched the historic event of astronaut, John Glenn, traveling into outer space in Friendship 7 spacecraft. I saw this momentous occurrence on my family's old black and white RCA television set. Prior to this experience I was always fascinated by science and scientists. I especially enjoyed learning about animals, nature and how scientists heal people. *The Astronaut's Window: Collection of Poems and Short Stories Celebrating Nature* is a book I wrote to acknowledge contributions of astronauts and scientists to humanity and the world.

As a teenager growing up in Pittsburgh coming from a lower middle-class family, I could not afford to visit the science museums and planetariums. I discovered through reading books, particularly science fiction, I could use my imagination to travel across the galaxy and universe. I frequently visited the local public library to read science journals and science fiction novels. Reading science fiction made the study of science exciting. It heightened my interest in scientific endeavors.

My science teachers influenced me greatly and gave me good guidance in my school years. They widened my perspective of how the world and universe work. They assisted me with perceiving the world like an astronaut in a universal way. I particularly believe that astronauts offer us an exceptional view of the world. From the windows of the astronauts' spacecrafts and space stations they help us to see the world in a distinct way. It is a special view. Learning about the astronaut's special view of seeing the Earth, myriad array of stars, planets and other space phenomena inspired me to write *The Astronaut's Window: Collection of Poems and Short Stories Celebrating Nature.*

Chapter 1

STARS WONDERFUL

The age of space exploration began in 1957 with the launch of Sputnik—Earth's first manmade satellite.

Now, humans have walked on the moon. Astronauts live and work on the International Space Station. We send probes to study the sun and distant planets. Rovers on Mars reveal crucial information about our neighboring planet.

How did we accomplish so much? Generations of curious people

I look at the night sky and see many extraordinary worlds…

THE ASTRONAUT'S WINDOW

I look at the night sky and see many extraordinary worlds…
I see Swift Mercury scampering across the evening sky,
    Playing hide and seek with sunset's orange-colored clouds.

I see Fragile Venus gliding gracefully across the heavens,
    She dresses elegantly in wreaths of comets and stars.

Bellicose Mars strides purposefully across the midnight sky,
    He wears vibrant red warrior garb,
    Fighting foes with mighty swings of the sword.

Lucky Jupiter arcs across the late night sky,
    Heralding good luck and vast fortune
    With many moon charms.

Austere Saturn slowly climbs the heavens,
    Promising toil and hardship in its passage.
    In return I gain fortitude and courage.

Thoughtful Uranus pours its universal knowledge,
    From an infinite celestial urn.
    I try to understand the complex ratios
    And algorithms of the universe.

Fluid Neptune's night passage is a swim,
    In a dark ocean of dream fragments and unfulfilled hopes.

Forgotten Pluto, no longer noble,
    Still demanding the harsh tribute of loyalty

And steadfastness from me in his night passage.

As I look through my astronaut's window,
    These wondrous worlds touch my being,
    Bringing me the wisdom of the universe in their night passage
    And I am thankful for their traverse across the night sky.

STARS WONDERFUL

Pegasus…
Phoenix…
Great Bear…
Lynx…
Fox…
Eagle…
Unite the night sky to tell us ancient stories.

Sun…
North Star…
Sirius…
Vega…
Orion's Belt…
Rigel…
Bright lights in the night sky are stars wonderful.

AMERICAN DISCOVERER

Tribute to Senator John H. Glenn, Jr., first American to orbit Earth.

He was a hero inside and outside of science,
Showing the world how innovative ideas are possible,
Sharing his scientific brilliance becoming a great humanitarian.

He was a hero within the science profession,
Hailed as a courageous pioneer of astronauts,
Showing the world scientific advancements.

Outside of science he was committed to a better America.
He was a friend of America and through politics helped people
By becoming a U.S. Senator and running for the presidency.

Heroic men and women take brave chances
To earnestly introduce new ideas,
Risking everything to validate these concepts.

New science ideas arise from brilliant people
Devising, nurturing and giving,
Their marvelous innovations to the world.

Sometimes new ideas in science
Move slowly through the world like a glacier,
Taking years to inch across rugged land.

There are times when new science ideas
Speed by quickly in a flash,
Like a pebble skipping across a clear pond.

One American especially needs commended

For bringing science innovations to light,
He was the first American to circle the Earth three times.

He was an Astronaut Discoverer,
Showing the world, innovative ideas are possible,
Being a hero inside and outside of science.

FIRST STEPS TO BEGIN

There had to be a first time in the universe,
>For stars to shine,
>Planets to rotate,
>Moons to glow
>And life to begin.

There had to be a first time in the universe,
>For comets to orbit,
>Asteroids to crumble,
>Meteorites to strike
>And solar systems to end.

When was the first time?
>Billions of years ago or in cosmic time,
>One minute that slipped past.

When was the first time,
>Love, empathy, hate and jealousy were felt?

What was the first cause to make
>These turbulent emotions happen?

As we consider the beginnings of the universe,
>With forces and energies of gravity, magnetism and electricity,
>Let us consider the beginnings of emotions
>And the forces and energies that cause feelings to appear.

NIGHT FLIGHT TO A SPACE SHUTTLE

Tribute to Dr. Guion S. Bluford, Jr., first Black American to travel into outer space.

I took a night flight to a space shuttle,
I was the first Black American astronaut to make this trip,
This astonishing, fantastic, amazing, astounding trip,
Yeah, I took a night flight to a space shuttle.

People say I am a special person for making this journey,
Like I am somehow different from humanity,
A god of sorts, only I am not,
I am just as human as you are.

I knew where I wanted to be,
From where I was now,
I knew I would have to work,
Beyond my imagination to go from now to there.

Obstacles became reasons to proceed,
I learned about physics, aeronautics, computers and business,
I joined the Air Force and was proud to be a military man,
Flying dangerous missions as a fighter pilot.

I moved through life always reaching above,
I received many awards for military, leadership and image.
I believe my greatest achievement, my best award
Was taking a night flight to a space shuttle.
Yeah, I took a night flight to a space shuttle.
I was glad to go and I was proud to take the flight.

DOCTOR ASTRONAUT

Tribute to Dr. Mae Jemison, first Black American woman to go into outer space.

Scientist Healer,
A Scientist who heals by mending the ravages of nature.
This Scientist has
    Compassion for curing life's hurts,
    Sympathy to appreciate nature's variety,
    Empathy in the rightness of living beings.

Scientist Astronaut,
A Scientist who explores through investigating universal laws.
This Scientist has
    Adventures by making time exciting,
    Interests in finding answers to nature's dilemmas,
    Patience to persevere in any direction.

Doctor Astronaut,
A Scientist who combines healing with exploring.
This Scientist has
    Vision to confront the future,
    Flexibility for science grows outwardly,
    Energy to learn beyond the current limits.

Doctor Astronaut,
    You lead science to new endeavors.
    You take science forward to new ways.

MY LUMINARY SLEIGH RIDE

A new spiral galaxy was discovered today,
Galaxy…as in a large group of stars greatly attracted to each other,
Spiral…as in a twirling, swirling mass of stars like our Milky Way,
How I longed to see this newly discovered space phenomenon,
A new galaxy flashing colors of red, blue and yellow.

Last night I secretly went to the grassy knoll by the cherry trees
The grass is thick and matted deeply
Springing back quickly from a touch,
The clear night sky revealed thousands of stars surrounding Earth,
I wanted to see the new galaxy so I created a luminary sleigh
By making a wish for a blue, shiny sleigh with silver rungs.

The sleigh appeared large enough to support
A trip of several million miles
And to store the many amenities necessary for my exciting journey,
Air, gravity, food, water, warm blankets
And scientific instruments to view spatial curiosities.
The single wish I uttered created the vehicle
From my need to see this universal wonder
And my need was greater in imagination
Than the reality of the galaxy's existence.

Five …four… three… two… one … Lift Off!
My luminary sleigh jerked upward leaving the grassy knoll behind
Launching easily with a single, imaginary jolt toward the night sky,
Speeding faster than light from Earth to the distant galaxy,
I clutched the arm rests tightly, glad that I had created them.

The sleigh careened through space passing galaxies seen before
Elliptical, oval almost circular masses, irregular multi-shaped figures
And even spiral galaxies but not the special one I wanted to see,
Within minutes the sleigh reached the particular distant galaxy
Hovering outside its enormous gravitational pull.

The sight of the spiral galaxy was worth the effort of my wish.
The grandiose view of the spinning, twisting rotation of millions of stars,
The varied colors sliding in and out of auburn, indigo and citron.
Traveling to this remote galaxy with its ordered form
Became my favorite drive in the luminary sleigh.

STAR FORERUNNER

Tribute to Dr. Bernard Anthony Harris, Jr.
First Black American astronaut to walk in outer space.

Dreaming as a child he glimpsed white-sparkling lights of stars,
Glittering through heavy, dark clouds,
At dusk in a purplish-blue sky,
He is fascinated by tiny sparks of light reaching out.

Making a promise to his future self he pledges,
When I am an adult I shall be an astronaut forerunner,
Traveling above Earth to view stars not covered by clouds,
Seeing how Earth looks from hundreds of miles above.

While growing up, he does not forsake his dream to travel into space,
By being dedicated, he is not deterred by life's ups and downs,
He takes necessary steps to complete his wish of journeying beyond
Through advanced schooling and rigorous training.

The young man carries his childhood's dream of traveling through space,
From the confines of Earth to the unlimited universe,
Through high school, college, medical school and astronaut training,
Soon his wish to voyage above Earth becomes a joyous possibility.

Embarking on a scientific voyage into the wondrous sky,
He circles Earth in a glorious spacecraft seeing stars not hidden by clouds.
Walking in space creating a special path in astronaut history,
Viewing Earth from a distant height as an astronaut forerunner.

MULTICOLORED PATCHWORK

Blue clouds rolling across the verdant-colored sky,
I stand beneath a cascade of green-metallic sheen.
My skin radiates and glistens,
Fluctuating with variations of dark and light.

In my world all people are multicolor,
A patchwork of dark and light hues.
We are proud of our variety,
Being one hue is a wrongness to us.

God created variety,
Our skin color reflects God's desire.
Other worlds scorn the kaleidoscope
Of colors and patterns we find enchanting.

Our differences do not divide us
They bring us closer together.
We are multicolor patchwork of many hues,
Fluctuating with variations dark and light.

Other worlds do not appreciate the need
And rightness of being different.
Our patchworks of varied colors
Make us see the variety of the universe.

CITY CONSTELLATIONS

At nighttime
Patterns of bright-yellow light
Shine from black-city buildings
Standing tall in regimented lines.

City myths
Tell of these glimmering lights
Shifting and forming magnificent shapes
Of giant-city heroes and mythical creatures.

Night gods
Newly arrived to Earth
From distant planets and stars
Mingle proudly with these city legends.

City folk
In their night dreams
Glimpse these special encounters
Capturing the unique gatherings in dance and song.

At dawn
The special occasions end
The building lights become ordinary
Night gods return to their planets and stars.

City people
Think only of metal and concrete
Work only to progress through daily life
Hope only for night gods and city legends to return.

Chapter 2

EARTH VIEW

Someday I will find this place where all my sorrow dies,
The place where flowers grow sweet and trees touch the sky...

A HAWK'S IMMORTALITY

Burnt-red tinged wings
Bound to a strong white-feathered body.
Rust-red domed head
Swivels left and right.
While sun-gold eyes
Search the horizon
Seeking prey to hunt.

Royal animal,
Worshipped by kings,
Praised as the greatest hunter.
Intelligent, far-seeing, all-knowing,
Fighter, conqueror, hunter.

Dark-grey talons stretching,
Reaching, sinking deep
Into a body cavity.
Pulling, ripping, taking a life,
Feasting for a moment on the hunter's victory.

Life's cycle spins again,
Reclaims the hunter's reason.
Life, death,
Life, death,
Beginning, ending,
The paradox of nature,
Beauty, power and strength
Determine life or death.

A GOLDEN BIRD

We will rise like a golden bird
    From the ashes of war,
We will shake off these worldly cares
    To live and love once more,
I am a soldier ready to fight for right,
You have to believe in my words
    Through these long nights.

Daybreak is coming soon over
    This cold, ravaged land,
Yes, we will see lives gone
    And weep for their passing,
I will grieve the hardest for the beauty
    We have lost,
My heart will ache with yours
    And wonder at the cost.

The golden bird will lead us
    On to another tomorrow,
Our days may still be filled
    With heartache, despair and sorrow,
We have freedom now to guide us
    Along the way,
We have freedom now
    We are no longer slaves.

ANANSI, THE SMALL ONE

Some people are afraid of me,
Some people run far away at the sight of me.

Why do you fear?
Why do you run?

I am small,
Only as big as your hand.

I spin beautiful webs,
That mystify scientists.

I am Anansi, the spider,
I have to be cunning to live
In a world so much bigger than I.

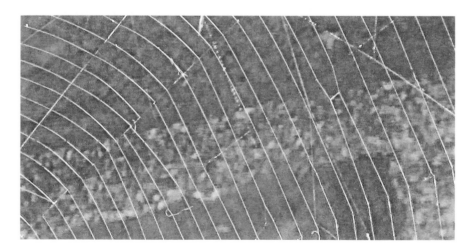

PIXIES — A GIFT FROM GRACE'S GARDEN

Fairy bells,
Pixie laughs,
Elf dreams
And this flower.

Wishing wells,
Golden bells,
Enchanted dreams
And pixies.

Magic hopes,
Magic thoughts,
Magic love unfolding.

Pixie world,
Pixie times,
Pixie life and sprite laughter,
All grow in Grace's garden.

May pixie love, laughter and hope grow always
In your life's happy garden.

FLOWERS

Fresh showers with,
Love and grace,
Open my heart to the,
Wonders and beauty of,
Each person's goodness,
Resting inside all souls.

Flowers mark the joyous and happy times of our lives.

COLORS

We shape colors to fit into our moral boxes,
No specific color is evil
Each color offers truth in different shadings.
A chair does not change by the color of paint it acquires,
A blue chair, a red chair, a yellow chair, all are equivalent,
Only how the chair is perceived changes.

We confuse the object that kills us with the coloring of the object,
The black coloring of the knife does not murder us,
The knife's metal point and wooden handle
Reaching our heart destroys us
And so we cry out black is evil when the knife not the color killed.
Apologies are not necessary,
Excuses are not necessary,
We confuse the object that kills us.

It must be remembered that words
Give only a guideline of color's truths,

Go deep within yellow,
Go deep within purple and find the truth,
The truth only words can partially give
And destroy the lies that pervade about the happiness of yellow,
We shape colors to fit into our moral boxes.

THE ALIEN IN LIFE

Complex chemicals with strange names
Of carbon, oxygen, nitrogen and hydrogen,
Can unite successfully to make a living being.

Yet the lives made up of these alien names,
Fight viciously and violently without remorse
To stay as far apart as is humanly possible.

What if each part that makes up life,
Does not cooperate, does not combine,
Does not unite and does not become one being?

Each part has valid reason for its actions,
Traced directly to honorable books of worship,
To legitimate state constitutions made by wise sages.

These ponderous documents all falsely show,
One life can be better than another life,
One life can be superior to another life.

The reasons given are astoundingly similar,
The reasons begin with the words "I cannot…"
The reasons end with the words "because you are not…"

These chemicals with strange names form bonds
Naturally without pressure without coercion,
Without requiring all the carbon to bond together.

Complex chemicals with alien names,
Can bond successfully to make a living being.
Can falsehoods be cast aside to unite a living world?

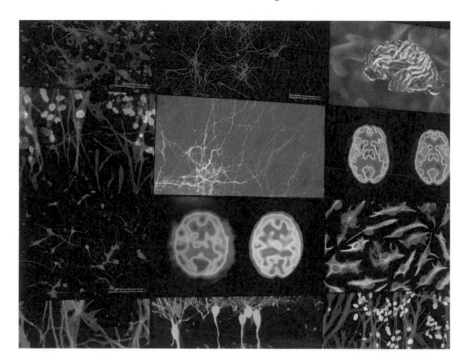

JOURNEY TO SALIBRI

There is a way I must walk to be who I am,
There is a road I must take that leads to the land,
The land of freedom.

And this place is where my hope lies,
This place is where all my sorrow dies
In the land of freedom,
The land of Salibri.

Flowers grow sweet there,
Trees touch the sky,
Rivers run swiftly there,
Valleys are wide and filled with joy.

Someday I am going to find Salibri,
Someday I will run where I am free,
Someday I will find this place where all my sorrow dies
Where flowers grow sweet and trees touch the sky.
The road to Salibri is what I must climb,
The land of Salibri is the place I am going to find.

Chapter 3

LEGENDS THROUGH TIME AND SPACE

To find all American heroes we need
to reach across space and time.

AMERICA'S REAL LEGENDS

America's myths and legends have many heroes
    Like Paul Bunyan and Johnny Appleseed.
America's myths and legends fail to take note
    Of heroes who look like me.

Dark skin wonders apparently did not loom large
    In America's past.
Buffoons and boogie men were how dark people
    Seemed to be cast.

In reality contributions by people of color
    Loom large.
In reality achievements by people of color
    Are ignored.

Crispus Attucks, Phyllis Wheatley, Sacagawea
    And Harriet Tubman,
Benjamin Banneker, Hiawatha, Pocahontas
    And Chief Seattle.

Named and unnamed these daily heroes
    Grace our land.
Their heritage, honor and freedom
    Make America stand.

All of America's real legends
    We can still find.
To find all American heroes we need
    To reach across space and time.

JOURNEY OF LIVING
Tribute to Dr. George Washington Carver.

I spent my life trying to help the farmer
By researching better ways to grow peanuts,
Sweet potatoes, soybeans and other crops.
I spent my life loving God
By respecting and honoring the farmer.

The old ones told me when I was born
My Mama laid me on the dirt floor of our shanty.
I took a small chunk of the brown dirt in my tiny hand,
I held the earth clod tightly to my chest
Laughing with sheer joy like I was holding my favorite toy.

That moment must have begun my love
For God's earth, sky, plants and animals.
Being a slave my Mama and I were soon separated,
But the deep love I have for this earth and God's deep love
For me stayed with me and held my life together.

Learning came natural to me
I reached eagerly for schooling,
I was the first Black student at Simpson College,
I was the first Black student at Iowa State College,
I was not afraid to say I did not know.

Somehow I knew God would show me
How to fill the empty parts of me
With knowing and caring about life.
My yearning to learn the wonder of life
And to understand nature spilled out of me.

God took me from a baby in Missouri
To a young man in Kansas, Iowa and then Alabama.
It was in Alabama where Booker T. Washington welcomed me.
Doctor Washington gave me the opportunity to lead
The prestigious Agriculture Department of Tuskegee Institute
And I stayed there the rest of my life.

Have any of you ever held a piece of the thick, red clay soil of Alabama?
Have any of you ever smelled the pungent odor of the red clay
As you crumpled it in your hand?
Have any of you tried to farm it?
Have to coax plants from its stubborn soil,
To feed a family and neighbors and America
From this contrary, red earth?
God made me hear and feel
The quiet plea of southern farmers
Trying to plant and grow crops to feed the hungry.
I spent my life listening and feeling this farmers' plea,
I spent my life helping the farmers' journey of living.

REVEALING FREEDOM'S TRUST
Tribute to Tuskegee University.

What is Booker T. Washington grasping
    With such a firm grip of determination?
What is Booker T. Washington pulling
    From the near broken back of a slave?
Is his right hand seizing the iron cover
    Of America's many injustices?
Can he really lift the heavy weight
    Of slavery's suffering and despair?

With a proud upright stance
    And a left hand open,
Washington stands unafraid facing
    A tumultuous future.
What is the correct way a person
    Fights society's injustices?
What is the right way to change
    Slavery's anguish to hope?

Critics shout loudly how Washington's
    Fight to appease is soft.
Others shout loudly how Washington's
    Caution slows the wheel of change.
Yet his slow pondering steps
    Yielded Tuskegee University,
A thriving monument of excellence
    In research and learning.

How can there be only one way

To combat the effects of slavery?
How can there be only one way
To remove the stench of this hideous disease?
Booker T. Washington was giving
One answer for curing the ravages of slavery,
Booker T. Washington was giving
One way to reveal freedom's trust.

AFRICAN LINEAGE FOUND

My African lineage is found
In the rhythmic cadence of my words.
My African lineage appears
In my profound respect for ancestors.

"Go Back To Africa!"
The strangers yelled at me.
"This is NOT your neighborhood!"
"This is NOT Africa where you can be free!"

But each day I walk in Africa,
My ebony body is where America and Africa meet.
My black, curly hair is a noble headdress
Worn by a regal tribal chief.

My flat, broad nose still smells
The wild jungle orchids' perfume.
My curved, effervescent smile
Glows like an African sky holding a crescent moon.

Like a supple African dancer,
I move through harsh city streets
With memories of dancing on lush savannahs,
I am where America and Africa meet.

THE VALIANT CIVIL RIGHTS LEADER
Tribute to Dr. W.E.B. Du Bois.

Doctor William Edward Burghardt Du Bois,
    Was a warrior battling social injustice,
Society's numerous wrongs were met
    By W.E.B. Du Bois remarkable intelligence.

A contemporary knight using protest as his fighting sword,
    Civil rights champion dismantling discrimination,
Political activist co-founding the Niagara movement and NAACP,
    Prolific commentator skillfully dissecting atrocities.

Revered by some and criticized by others as being
    Combative, demanding too much too soon, unrealistic,
W.E.B. Du Bois believed that segregation, discrimination and racism
    Can end through protest, lobbying and legal forays.

Doctor William Edward Burghardt Du Bois,
    Was indeed a great warrior battling social injustice,
Society's numerous destructive wrongs were met
    By the extraordinary intelligence of W.E.B. Du Bois.

THE MISSION OF EDUCATION
Tribute to Dr. Mary McLeod Bethune.

Mary McLeod Bethune worked tirelessly
    Opening up education for Black people,
Doctor Bethune was vital in giving
    Black people chances to succeed,
Doctor Bethune taught us that
    Education quiets the many fears,
Doctor Bethune showed us that
    Education allows us to grow.

We come into this world
    With basic instincts,
We come into this world
    Able to smile, weep and cry,
We come into this world
    With an abundance of feelings,
We come into this world
    Not understanding how or why.

Mary McLeod Bethune's accomplishments were many,
    –Teaching her 16 brothers and sisters how to read,
    –Teaching in mission schools throughout the South,
    –Appointed by FDR to direct youth training programs,
    –National Council of Negro Women founder,
    –Vice President of the NAACP,
    –Representative to the start of the United Nations,
    –Encouraged FDR to tackle employment discrimination and
    –Founder and President of Bethune-Cookman College
Were only some of Mary McLeod Bethune's achievements.

Educating the world to see
    That equal and fair are good.
Educating the world to believe
    We can reason and have civility.
Educating the world to understand
    When Black people prosper, America prospers.
Educating the world to understand
    Mary McLeod Bethune's works can be reality.

A SOLDIER'S STRENGTH TO FIGHT

How do I find the strength to fight?
I have fought hard through the long night.
My enemy has all the might
He is winning and that is not right.

My enemy has done many evil things,
He hunted down and murdered our valiant king.
Then he unfolded his massive wings
Flew into our town destroying, destroying.

I have the will to succeed,
Time to rest is all I need.
Heal my wounds and I will soon lead
And make our dying world be free.

I will lay by this jasper stone
And dream of an empty throne.
In these dreams I will mourn
For my father leaving me alone.

How do I find the strength to fight?
I will find inner courage to fight for what is right.

MONOLITHIC GUILT

Faithfully each day we climb our monolith of guilt,
Uneven crevices filled with self-doubt and uncertainty
Shape the monument into an odd misshapen body.
By habit we find the same handholds and footholds
We foolishly grasped yesterday.

Our climb begins with a memory from a long time ago
Where we selfishly sought and obtained a needless desire.
Trampling over others in our rush to fulfill this whim,
Pausing to wince at the still raw wound exposing
Our deep ingratitude.

Moving on we climb over more recent memories,
Speaking deliberately with cruelty, finding ways to be vengeful,
Ignoring other people's desperate needs to be alive.
Until finally we arrive at the top
Of our important monument.

Climbing the same route every day
We learn nothing in our pilgrimage.
We prefer the tedious barren process,
Finding it easier to navigate the painful slopes
Than to abandon our monolith of guilt.

BROTHERS AND SISTERS LET US TAKE OUR FREEDOM NOW

Brothers and Sisters, Let Us Take Our Freedom Now!
Let us shape it wide and loud,
Let it sound throughout our country,
Let freedom's presence make us proud.

We have marched too many steps,
Carried too many signs,
We have sung too many protest songs
Pleading our cause too many times.

For the right to own land,
The right to have equal wages,
The right to vote our conscience,
The right to break from these cages.

We are no longer trapped
Afraid to do what we desire.
We can take our inner strength
Leave society's degrading mire.

Brothers and Sisters Let Us Take Our Freedom Now!
Let us chose the things we need
To win our independence,
To create a fair and just society.

Let Us Take Our Freedom Now!

51

Chapter 4

LEFTOVER TIME: PAZA'S ADVENTURE STORY

"Would you like to buy Leftover Time?

LEFTOVER TIME: PAZA'S ADVENTURE STORY

Paza Martin glanced at the wall clock hanging above the doorway of her cramped library office. The clock showed seven o'clock. She shifted her eyes to the piles of unevenly stacked books covering her desktop in the middle of the room and placed on metal tables lined along the sides of the office. She sighed heavily.

"Two more slow, unforgiving hours," Paza muttered.

She sighed again and flexed her stiff shoulder muscles trying to relieve the tightness she felt in the middle of her neck from sorting books for hours. This evening was the last day the public library in Philadelphia paid overtime to employees working on the annual book inventory.

Needing money, Paza took as many hours of overtime as she could. She had recently graduated from college with a liberal arts degree accumulating a large amount of student loan debt she needed to repay. Paza's family could not afford to pay for her college expenses. Her parents held low-income jobs in Harrisburg with her mother being a waitress and her father driving a taxicab to care for five children. Paza had excelled in high school. Due to her high grades she obtained a partial scholarship to attend college in Philadelphia. The scholarship money was not sufficient so she had to acquire student loans to finish college.

When Paza moved from Harrisburg to Philadelphia four years ago, her mother, Isadora, made her daughter promise to keep her dream going.

"Promise me, Paza" Isadora requested of her eldest child. Isadora had come to the bus station to see her daughter depart for college.

"Promise me," Isadora further implored, "you'll follow your dream to become a writer when you graduate from college. I believe you have great talent."

"I promise," replied Paza hugging her mother goodbye as tears tumbled from her oval, black eyes. Paza boarded the bus to Philadelphia waving goodbye to her mother reluctant to leave her loving family.

Having the library clerk job also helped Paza to be around books. She had lacked experience to obtain her first-choice of working in a publishing house. She was thankful to be working in a library hoping it would inspire her to write a novel.

Each morning since she graduated six months ago, Paza surveyed herself in the dingy bathroom mirror of her sparsely furnished one-room apartment she now called home. She saw a petite, dark-brown woman with black tight-curly hair, searching eyes, wide nose and full lips. Her mother said Paza was attractive and sometimes Paza agreed with this assessment when she felt confident.

Looking at her blurred image in the mirror Paza said, "I will write today."

In the evening when Paza returned to her dark apartment, she switched on the light to eat her frugal dinner of baked beans and hot dogs or macaroni and cheese. Then she went to bed. She was too tired from working twelve hours a day handling books to take time to be creative and write.

"Tomorrow, I will write," she said to herself and of course she did not.

As days passed, Paza made more excuses for not writing. She needed to take more classes to hone her writing skills. She had no ideas for writing a book. Who would read the books she wrote?

This morning, she frowned when she looked at herself in the mirror and this time she said, "You can't write." She felt relieved after making the revelation and less guilty for not honoring the promise to her mother.

Paza glanced at the clock again and the time was seven fifteen. She yawned, blinked her eyes and rubbed her face. She tried to stay awake but her head slowly drooped to a space she had cleared on her desk top. Paza fell asleep and began dreaming of living in the future.

"Gracious lady, would you like to buy Leftover Time?" Striver Paza found herself sitting at a table in the candlelit dining room of Slair Merchant Inn located at the outer region of the city, Teres, in the Northern Continent on the planet Hilo 2.

Striver Paza raised her head at the interruption. She was dozing, slouched in her chair, with her head nodding over her partially eaten dinner of vegetable stew and bread. She sat alone at the dining table where tapestries of old Earth landscapes were barely visible hanging on the wall behind her seat.

It was almost midnight. Paza had returned an hour ago by electric carriage from the Southern Continent where she had been working on her first Striver assignment. At 21 years old, she was a newly appointed Striver Counselor, an advisor and confident of kings and nobility. Five arduous years had been spent studying magic, diplomacy and science to become one of the elite group of Strivers on planet Hilo 2.

Paza came directly to Slair Merchant Inn, her home base in Teres. She booked her usual room and drifted into the dining room to eat a midnight meal. Exhausted from eight hours traveling in a crowded electric carriage, Paza ate little of the stew filled with tender chunks of spring vegetables.

She was also tired from rehashing to herself every detail of how she handled her first assignment. She had failed miserably by badly maneuvering the negotiations of a land deal between two feuding states. It was necessary for a senior Striver to finalize the negotiations. Throughout the trip back to Teres, her throat ached from holding back tears. When Paza arrived at Slair Merchant Inn, she relaxed a small amount from the Inn's warm surroundings of tapestry-covered walls and soft, bulky furniture.

Paza focused her eyes at the sound and saw a Rall standing at her table. She had not heard the being approach. Her large, black eyes widened and she sat more erect. She had never seen a Rall but she recognized the race. They had characteristics of being about 180 centimeters tall, thin, having lavender eyes, dark-brown skin tone and six-fingered hands. These features matched the humanoid being in front of her. He looked to be about 25 or 30 years old.

"Gracious lady, would you like to buy Leftover Time?" asked the being again in a firmer and louder tone.

"Huh…Of course…I mean, maybe?" responded Striver Paza somewhat groggy.

She winced at stumbling over her words. She was supposed to be well-trained in the graces of society as a Striver. She should not act like an unschooled child unable to speak properly.

The being slid uninvited into the wooden chair across from Paza.

"I am Zir, a lowly Slair Merchant," he said as he reached into the front pocket of the dark-blue tunic he was wearing with matching dark-blue pants and boots of the same color. He retrieved a reddish-brown plain metal box about 15 centimeters long The box glowed with a yellow gleam. Zir placed the box in the middle of the table.

Paza pushed the cold bowl of half-eaten stew to the side and reached for the box interested in learning its contents. Zir quickly placed his hands on top of the box blocking her reach. Frustrated by Zir's actions, Paza drew her hands back, placed them in her lap and gazed openly at the Rall for several seconds.

Paza, now fully awake, was irritated by the boldness of the Rall in blocking her from picking up the box and pretending to be a Slair Merchant. She knew much about the indigenous Rall people of Hilo 2 having written a research paper on their culture as an apprentice Striver.

She recalled pertinent facts from her paper. Ralls were present on Hilo 2 when the Earth colonists arrived ten thousand years ago from a polluted, overpopulated Earth situated across the Milky Way Galaxy. The Ralls had been relegated to the inhospitable Western Continent by the encroaching Earth colonists. The two remaining continents, connected by a narrow land bridge, were called Northern and Southern. These continents surrounded by oceans had temperate climates with moderate terrains and were eagerly settled by the colonists.

Ralls did not fly spaceships like Slair Merchants who sold expensive oddities to wealthy clients from planet to planet. Ralls did not use electricity in their houses as the wealthier colonists did. The Rall homes were simple and unadorned places. They kept to themselves and occassionlly interacted with the colonists. So why the scam? Paza spoke her concerns out loud.

Paza said, "Look Mister, my name is Striver Paza."

Zir sat up straighter and leaned forward towards Paza.

Paza added, "You are not a Slair Merchant. You are a Rall. So why the masquerade of being a Slair Merchant selling something phony called Leftover Time?"

Skepticism of Zir could be seen in Paza's oval dark-brown face from the tilt of her head with a cap of black tightly curled hair to the black-arched eyebrows over her large, black eyes. Her wide lips were turned down and her crinkled, broad nose seemed to be smelling a putrid smell. Her petite, 155 centimeters body was clothed in her favorite outfit of a black floor-length shift with black boots. Paza stared at Zir waiting for an explanation.

"Striver Paza, you are right. I am a Rall," Zir replied, "I am not selling a false item to you. I come to you for help and an alliance. How can I make you believe me?"

Zir was doing badly in explaining his need for a Striver. He was finding that he liked this Striver's sincerity and directness. Paza's presence seemed to affect him in a remarkable way. He was not being his usual articulate self. He seemed clumsy with his words. He took his hands away from the gleaming box and folded them at the edge of the table. He waited silently for Striver Paza's next move letting her speak without further correcting her.

"I cannot accept your claims so easily," responded Paza. "You came to me based on a lie of being something you were not. So why should I believe you now?"

She was going to be thorough this time. She was not going to believe everything that was said as she did on her first assignment. She would prove every comment if necessary.

The conversation was moving into areas where Paza had no expertise. How could there be an alliance with the Ralls? What aid is needed? What was Leftover Time? Paza decided she needed the expertise of another Striver. She had waited far too long to obtain expert Striver advice on her first assignment according to the Master Striver who had helped her. His final solution was to have the two countries share the disputed land by both countries farming the land and equally dividing the profits from the sale of the produce. Paza, lacking experience, embraced the bickering, petty dislikes of the two countries without giving a viable solution.

"I need to get some other Striver here, Zir," cautioned Paza, "I have to be honest I do not know enough to agree to an alliance or offer help."

These words were difficult to say. Paza was proud of her intelligence and independence. She was humiliated by her need for help. Humility did not make Paza stronger. It made her doubt her analytical skills.

Paza came from a farming village, Mele, 500 kilometers south of Teres. She was raised in poverty by a single-mother who had passed away five years ago. To overcome the shame of being poor and to demonstrate her exceptional intelligence and skills, she became a Striver.

From her table, Paza searched the room filled with circular dining tables covered with white-linen tablecloths. There were no customers at the dining tables. The flickering candlelight from white-frosted candle holders on the dining tables was augmented by electric floor lamps placed in corners.

Paza was hunting for the tall, thin figure of blond-haired Master Striver Zachery, the owner of Slair Merchant Inn and her mentor. A direct descendent of the first Earth colonists like Paza, Zachery was a knowledgeable, highly respected Striver Counselor. Paza knew she was fortunate to apprentice under Zachery. His easy-going nature made the difficult path to becoming a Striver tolerable. Then Paza remembered that Master Striver Zachery was out-of-town on Striver business.

She looked for his eldest daughter, Destani, a blond medium-sized copy of her father. Paza had apprenticed with Destani in training to be a Striver. They had become close friends. Paza had become a Counselor Striver while Destani chose to become a Healer Striver.

When Paza entered the Inn she had seen Destani wearing a green-satin shift with a gold waistband supervising the Inn's staff. The two Strivers only had time for a quick "hello" and "how are you?" Paza located Destani sitting in a lighted corner at a massive oak table attached to the back wall of the room. She was sifting through various papers on the table and laughing.

Paza noticed that another newly appointed Striver was sitting at the back table. Paza grimaced when she saw it was Kaiden. He must have entered the dining room while she was dozing.

Twenty-one year old Kaiden was also a descendant of Earth colonists and was an astute Scientist Striver, the third type of Striver. He was handsome, about 180 centimeters tall, muscular and bronze with curly-auburn hair complementing his pleasing light-brown eyes. He was wearing grey walking boots and a grey traveling outfit of tunic and pants made from the wool of the indigenous whirr animal. This animal resembled an Earth camel with wings. It was a sturdy animal that could negotiate well the varied terrains of Hilo 2.

This Striver enjoyed the analysis part of science. His parents who were both Scientist Strivers often found Kaiden carried his analytical abilities to the extreme. He would ask a multitude of questions to intensely probe any situation heedless of how he was affecting a person.

During Paza's apprenticeship Kaiden's overt attention towards her had been a nuisance. He always managed to team with Paza in group assignments, give her numerous gifts of candy and flowers and asked her several times on a date. Paza politely and sternly returned the gifts and refused dates with him.

At the back of the room, Kaiden seemed engrossed with entertaining Destani with magic tricks. Both Strivers were laughing at the antics of Kaiden's tricks. He had created miniature Earth landscapes hanging in mid-air with herds of Earth animals moving across the land. He produced African savannahs with lions roaming under a hot sun, forests with leaves colored orange and bright-red had families of deer winding through the trees and white glaciers had polar bears lumbering past. Muffled sounds of animals growling were faintly heard.

Paza waved for Destani and Kaiden to come to her table. They saw her wave, the landscapes disappeared and the two moved to Paza's table. Kaiden brought two additional wooden chairs for Destani and himself. He placed the chairs on either side of the table and the two Strivers sat down.

Destani smiled at Paza while twisting her shoulder length blond hair and asked, "What is happening, Paza?"

Paza swept her right hand in the air to include Zir and the gleaming box on the table.

"We have a problem and I believe it is serious," answered Paza, "This person is called Zir and he has requested an alliance with Ralls and our help."

"What kind of help?" asked Destani no longer smiling. She stopped fidgeting with her hair and looked at her friend.

"I think it is best if Zir spoke and explained why he needs help," responded Paza.

Zir lifted his eyes from where they had been focused on his folded hands. He spoke with his words flowing rapidly emphasizing the importance of the matter.

"Striver Paza, you are a beautiful gracious lady to have your friends hear my story."

Paza looked sidelong at Zir at the mention of being beautiful. She had not expected that reaction from him.

While Zir told his story, the three young Strivers listened intently letting Zir unfold his tale.

"I am a Rall priest from the Western Continent," began Zir. "I came to Teres this morning on a Rall fishing boat crossing the ocean between the Western and Northern Continents. I asked the sailors at the dock where to find Strivers. They told me to come to this Inn."

Zir continued, "You see, we kept much of our culture hidden from the colonists especially the magic and symbols. Every one hundred years, six Rall priests take their temporal boxes containing Leftover Time, like this box on the table, to open the Time Haven Cave with the magic energy from the boxes."

"What is Leftover Time?" interjected Kaiden as he gazed at the gleaming box on the table. He longed to open the box and see how this object appeared.

"Leftover Time is the time between speaking a magic spell and the actual completion of the spell," responded Zir. "The few seconds to few minutes during this time period contains potent unused power. This power can be collected at the end of the spell using specialized instruments."

Zir proceeded with his tale, "The Leftover Time changes the yellow, stone slab sealing the cave's opening to a sheer curtain. The priests push the curtain aside to enter the cave. After entering the cave we perform a restoring ceremony to renew the energy of this planet you call Hilo 2. If the ceremony is not completed Hilo 2 will be ravaged by massive earthquakes and torrential weather destroying both Rall people and the colonists."

"Two weeks ago, the priests, and I was one of them, tried to open the door to the Time Haven Cave. The door did not open. We used the same magic ceremony that has been performed for thousands of years. We know it is the same because the ceremony was carried down to us through our oral traditions and this ceremony was meticulously described in ancient texts."

"We tried other methods to open the door such as blasting powder and fire. They did not work. The Ralls governing council and the priests agreed that we needed to seek outside help. We do not have much time left before the planet-wide cataclysm happens. Even on the Western Continent we have heard of the magic power and great knowledge of Strivers. I come in the name of the Rall nation to plead with the Strivers for help."

Zir found a ready audience to his tale in the three Strivers. The enormity of Zir's circumstances greatly troubled the three listeners. They were deeply engrossed in Zir's tale and being inexperienced failed to be on alert.

While Zir was describing the devastation the planet could experience if no restoring ceremony was performed, the Inn's ornately, carved wooden door slowly swung inwardly open. A man dressed like Zir entered. He had similar features to Zir of a thin body, dark-brown skin tone, lavender eyes and six-fingered hands. He was much older than Zir with grey hair showing under the hood he was wearing. The man turned his head right and left as if he was searching for someone in the dimly lit room.

Within a few seconds of entering the room, Zir noticed the newcomer and instantly recognized who the intruder was. He was a Rall priest called Till who had vocally disagreed with Zir seeking outside help with the Strivers. Agitated by his presence, Zir jumped out of his seat knocking his chair backwards.

"Till, what are you doing here?" shouted Zir at the newcomer. "The people decided we must seek outside help. Why are you fighting the Council's decision?"

Till did not answer. He had followed Zir from the Western Continent. He was a Rall priest who disagreed with the Council's decision to seek the help of Strivers. The Council could not believe that a Rall would have the insolence to deliberately prevent the Time Haven Cave door from opening. Only Till did have the gall to interfere with the ceremony.

Disillusioned and discouraged by being overlooked for younger, less talented priests, Till frequently roamed alone on the rugged peaks and the ravaged valleys of the Spider Mountain Range located on the Western Continent. In his aimless wandering, he happened to find a lost cave containing ancient Rall magic objects and books. He did not know or understand how all of these significant objects came to be in the cave. He told no one of his find. He would frequently return secretly to the cavern sifting through the objects and studying the books to increase his knowledge of magic.

Till was waiting for the opportunity to become the sole leader of the Rall society. Before the Earth colonists, the Ralls had kings and queens with their history full of descriptions of these historic legends. Now they were loosely governed by a six-member council elected by adult Ralls.

What motivates a being to do anything including destroying others to be the ultimate leader? This driving nature has been labeled as greed, jealousy, selfishness and hate. Till may have harbored all of these feelings or just one of them if the compulsion was strong enough. What were the reasons for Till's intense longing to be superior? It may have been his unkind childhood of being an orphan and raised by disinterested relatives or being constantly overlooked and disregarded whether intentionally or not or having loveless relationships. Whatever the reasons, Till needed to be the King of the Rall race at any cost.

Till used an enchanted stone from his cache of magical objects to secretly drain the magic energy from the door sealing the Time Haven Cave. The priests' six temporal boxes of Leftover Time could not provide enough energy to open the sealed door. Till planned to open the door to Time Haven Cave in due time by restoring the magic to the door and thus becoming the savior of the Rall people. Then he would be praised and eventually proclaimed King.

First Till had to stop the outside help. He had to hinder the Strivers by frightening them. Till had to show them magic beyond what was generally heard about the Strivers' paltry attempts at casting spells.

Till stood at the Inn's door not speaking. He glared at Zir with intense hatred. With Till's right six-fingered hand, he reached into a pocket in the front of his dark-blue tunic. He pulled out a round, green crystal about 10 centimeters in diameter. He held the crystal close to his face whispering a few indistinct words. Then he threw the crystal on the floor shattering it into many pieces. The shattered crystal released a thick plume of indigo-colored smoke with a malignant stench. The smell was of decaying cadavers and rotting herbage.

Till pointed his right hand at Zir. The dark-blue smoke congealed into a fluid mass with grotesque, howling figures swirling in it. The smoke rapidly moved to Paza's table enveloping the four people sitting there. The overpowering smell made them double-over coughing and gasping. While the figures grew hands and fingers that grabbed at the young people shoving and pushing them. The four people tried to protect themselves by grappling with the smoke. They were unsuccessful. They sunk to the floor of the Inn gasping to breathe.

In the midst of the confusion Till approached Paza's table and removed the temporal box of Leftover Time from the table. Till wanted to deplete Zir of any magic. The poisonous smoke and swirling figures did not affect Till because he directed these maleficent objects only to affect the four young people and not himself.

Till turned to exit the dining room. Zir stood up and tried to follow Till. However Zir succumbed to coughing fits and collapsed in his chair.

The Strivers Guild has many sayings to cover the inconsistencies of life. The Strivers use these sayings as guides to understand the incongruities of life. One of their sayings is "When the Gods have deserted you, faith will guide you."

At that moment Master Striver Zachery chose to walk through the Inn's door. His lean six-foot frame collided with fleeing Till making both of them collapse on the floor. Till dropped the temporal box he had stolen from Paza's table. Till managed to disentangle himself and rush out of the Inn's door leaving the box behind.

Zachery immediately recognized the glowing box as an important object and quickly picked it up from the floor. Zachery allowed Till to leave the scene directing his attention to saving the four people caught in a magic maelstrom. Zachery swiftly assessed the enormity of the situation with four people enveloped in poisonous smoke and evil figures pommeling their prone bodies. The sounds from the figures shrieking and the groans of the four victims were unnerving.

Zachery shouted at the Strivers to join hands. He told them to say the Poem of Well-Being out loud. He began chanting the Poem of Well-Being also.

Kaiden, barely conscious, followed Zachery's shouted command by seizing the hands of Destani and Paza. The three young Strivers' weakly grasped hands creating a fulcrum of power. With eyes tearing and in-between coughs they chanted the potent Poem of Well-Being hoping to flood the Inn with an alternate power to the one released by the crystal.

After a few minutes the air cleared with the strange figures disappearing. The four people recovered. They could breathe easily. The bruises from the pommeling were instantly healed.

All four began talking at the same time checking with each other to make certain they were not hurt. They persisted talking on top of each other thanking Zachery for his help and trying to explain what had happened.

Zachery silenced the excited group by speaking loudly to be quiet. When the group became more composed, Zachery suggested they sit at another table in the dining room and start at the beginning of the story. As everyone moved to a distant table across the room, Zachery made certain the Inn's door was locked. He suggested that Destani get tea from the kitchen to refresh everyone which she did.

When everyone was seated at the table sipping their cups of spiced tea with the temporal box in the center of the table, Paza took the initiative. She explained the reason Zir was at the Inn. She emphasized the danger Hilo 2 faced with the inability to complete the restoring ceremony.

Zir filled in the parts of the story Paza did not know. He told about Till's dislike for outsider help. However, Zir could not explain how Till could become so powerful.

"I did not realize the depth of Till's hatred," said Zir. "A few...a very few of us believe that the colonists desecrated our planet and should be destroyed."

"And what do you believe?" asked Zachery trying to discern where Zir's allegiance was.

"I believe saving Hilo 2 is the most important matter," answered Zir.

"Then let us figure out how to save our planet," said Zachery.

They discussed the complex situation through the rest of night. Kaiden asked numerous questions fleshing out the matter to find a solution. When the morning rays of Hilo 2's sun spilled through the three large windows facing the city streets in the dining room, they were still debating what to do.

Paza remained quiet throughout the discussion. She hesitated to present her suggestion. Would the others think it was too simple? Then she remembered a Striver saying, "Find the easy way and you find the best way".

During a lull in the conversation, Paza said, "My suggestion is to collect Leftover Time from magic spells conducted by Strivers. The Leftover Time would probably be different and may stimulate the door to the Time Haven Cave to open."

Everyone agreed Paza's suggestion was the best and they decided to follow her idea. Paza's confidence in herself returned at the acceptance of her idea. Mistakes were part of living she thought. Her first assignment was a mistake and she needed to grow from her mistake not hide from it.

Paza turned to Zir and asked, "If we have to create Leftover Time, can we see what it looks like?"

"Sure," Zir replied. "Just do not touch it because it is very potent."

Zir opened his temporal box and showed the group Leftover Time. It resembled thin sheets of silver paper that glistened. Zir speculated that Striver Leftover Time may be another color.

None of the Strivers had visited the Western Continent and they relied on Zir to obtain passage for them.

"There is a Rall fishing boat waiting to return me to the Western Continent," said Zir when the discussion pivoted to transportation. "I am quite sure the boat is large enough to fit all of us."

"I think I better stay here," said Master Striver Zachery. "What happens if our scheme fails?"

Paza grimaced at hearing about the possible failure of her plan. Zachery noticed her reaction and quickly added.

"I am quite sure the plan will succeed," reassured Zachery. "However, it is always good to have a second plan. I will stay here and wait for two weeks since that is the time Zir estimates we have. If there is an upheaval then I can gather other Strivers to see if we can use our magic to find a solution."

Zachery gave each of the four a talisman made of gold from the planet Earth shaped like a miniature whirr animal with obscure lines written on it.

"Keep this object close to you," Zachery instructed. "It will strengthen your power by focusing your magic into a stronger beam of energy."

While packing for the trip to the Western Continent, Zachery approached Zir as the magician was helping Paza lock her traveling bag.

"Zir, we pride ourselves as Strivers that we know so much about Hilo 2," Zachery began. "Now I see how little we do know."

"Would you consider becoming a Striver?" Zachery asked. "You probably can teach us more than we can teach you. We need the Rall understanding of magic."

Zir was startled at the request to become a Striver. There were so many hurdles to climb over such as leaving the Rall people, living among strangers and being the different person.

"I need time to think it over," Zir replied.

Zachery said, "It usually takes five years to become a Striver but I think you could make it in much less time. Think about it and let me know when you return. I have faith that you will come back."

Communication was not certain on the Western Continent. They were unsure about communicating over long distances through person to person thought projection or even using electric callers. Till could have placed a damper over parts of the Western Continent making communication difficult.

Two weeks later, there were no violent upheavals in Teres. The sun shone brightly. People walked in the morning through tumultuous city streets lined on both sides with tall maple, oak and beech trees. The sidewalks near the Inn were full of pedestrians window-shopping. The streets were clogged with the usual traffic of colorful carriages pulled by whirr animals containing loads of furniture and clothing. Electric carts rolled by carrying important citizens.

A month passed. There were no earthquakes, no hurricanes and no erupting volcanoes. One evening, after Zachery locked the Inn door, he heard a knock on the door. Zachery opened the door to find Paza, Zir, Destani and Kaiden standing on the doorstep.

Paza blurted out the news before anyone could speak, "We did it. We were successful."

Zir added, "Striver Leftover Time worked."

Master Striver Zachery motioned for the four people to enter the Inn. Zir continued explaining what happened as Zachery took them into the kitchen to reheat a pot of stew for them.

"We were able to collect Striver Leftover Time from magic spells conducted by the three Strivers using the whirr animal talismans. We placed the golden-colored Leftover Time into six temporal boxes and applied it to the Time Haven Cave door."

Kaiden interrupted the account excited at being involved in this great magical endeavor, "The yellow, stone door to the cave changed to a yellow, soft-sheer curtain that could easily be moved away. Six priests including Zir entered the cave and conducted the restoring ceremony. The priests were able to restore the planet preventing a cataclysm."

Zir remained in Teres and decided to become a Striver. He felt responsible for Till's actions and the fact that Till with his powerful magical ability was not found. Zir found it interesting to study Striver science and magic. Zir's intelligence, dedication and being a highly skilled magician speeded the process of becoming a Striver.

Within two years, Zir was officially made a Striver. He stayed in Teres a few months to give Paza time to decide about their future. Zir also found it exciting to develop a relationship with Paza. Would they remain together or separate?

"You cannot keep saying give me more time to think," Zir admonished Paza when he asked her for the third time to marry him. "I have to return to the Western Continent and find Till or at least find out what happened to him."

"I spoke to a Rall fisherman this morning and he said there is a rumor about a fanatic being seen in the distant parts of the Spider Mountain Range by some isolated Rall herdsmen of whirr animals. This deranged man talks to the sky and rocks. By waving his hands he brings lightning and hail from the sky. This man needs to be investigated. He could be Till."

Paza had secretly welcomed Zir's student status as a Striver. She found she enjoyed his company. She liked sparring with him over how to conduct the more complicated spells. She had to admit that she loved him. Life would be empty without Zir.

"Okay, Zir," responded Paza. "I will marry you. Someone has to help you find Till. Anyway, I would miss helping you with your magic spells."

Paza and Zir were married with Master Striver Zachery's blessings shortly after Zir became a Striver. They moved to the Western Continent to live and work together as a team. Zir became a Counselor

Striver like his wife. The future was bright, exciting and dangerous for Striver Paza and Striver Zir. With their love and their Striver talents and abilities they should be able to surmount their future be it difficult, challenging or easy.

Their two friends, Destani and Kaiden, had married two years earlier. They promised to visit the newlyweds on the Western Continent very soon. They also wanted to help locate and curb Till.

There were many unanswered questions about Till that needed to be solved. Where was Till? How did his magic become so powerful? What were his plans and did he intend to destroy the colonists and the Rall race?

In her dream Paza heard a voice say, "Hey Paza, wake up!"

Waking up surprised to see the other library clerk, Jane, standing beside the desk, Paza suddenly sat upright in her chair. She looked at the clock that now read eight thirty. She realized she had been dreaming this story.

"Oh no, it was only a dream. Now I have the story for my novel," said Paza delighted with her exciting dream. Her wages would be docked for sleeping on the job and she would receive a stiff reprimand. However, the negative repercussions seemed secondary to Paza having a story to write and the wish to write her dream adventure. As soon as Paza arrived home to her one-room apartment, she began drafting her dream story naming it *Leftover Time*.

Chapter 5

NO GARDEN THIS SEASON: GRANDMA LILY'S GARDEN STORY

"Colorful flowers were planted strategically in the garden…making it look like a giant butterfly about to take flight."

NO GARDEN THIS SEASON: GRANDMA LILY'S GARDEN STORY

The bees were humming in Grandma Lily Johnson's garden as they flew from one inviting flower to another on a bright summer day in early June. Grandma Lily working alone on her knees in her garden was humming the same tune as the bees. The rising sun cast a soft light on the red brick wall surrounding the garden on three sides. As the warm rays from the sun touched Grandma Lily's face she felt more energetic.

"Today is a beautiful Saturday morning," said Grandma Lily to herself as she was pulling weeds. "I hope my son comes over soon with his family." Grandma Lily then paused and stood to admire her garden handiwork.

Colorful flowers were planted strategically in the garden behind Grandma Lily's three-story blue, colonial house in Philadelphia making it look like a giant butterfly about to take flight. Golden marigolds streaked with auburn lines were planted in the southwestern corner of the garden. Pink and purple impatiens grew in the shady, southeastern corner of the garden under the branches of the neighbor's imposing sugar maple tree. In the northwest section of the garden silver and yellow snapdragons were planted and in the northeast section pink geraniums grew in abundance.

At that moment, Grandma Lily's 35-year-old son, Edward Johnson, walked through the black wrought iron gate leading into the garden from the southern side of the red brick fence. His house was next door on the western side of the garden.

"Mama," said Edward as he tapped his mother on her shoulder to get his mother's attention. He was six-feet tall with pleasing features. He had a round dark-brown face with no beard or moustache. He had short, black hair, round-brown eyes, a wide nose and large mouth that frequently smiled. He was of medium build and was wearing his second-best green suit.

Grandma Lily was standing beside the marigolds where she had been pulling weeds from around the hardy plants. She was petite being five-feet tall, plump and had tightly curled-gray hair. She had a round dark-brown face, oval-black eyes, a broad nose, full lips and wore wire-rimmed glasses. Grandma Lily was wearing her gardening clothes consisting of a scruffy, colorless shirt and worn khaki pants.

"My job is requiring that I go to California for two weeks," continued Edward. "My company needs me to create special instructions on computers out there to make them work more efficiently."

"I thought it would be nice to have Felicity come with me. I thought I could put her traveling expense on my new credit card," said Edward.

"Great," said Grandma Lily. She often thought how her son and his wife, Felicity, deserved a vacation from their hard-working routine. They could not afford the expense of a vacation with two school-age children, Edward working a moderately paying job and Felicity working part-time as a supermarket cashier.

Almost as if Grandma Lily could read her son's mind she said out loud what Edward was thinking.

"I'd enjoy having my grandchildren. Why don't you have Jessie and Lamar stay with me?" asked Grandma Lily.

Grandma Lily added, "Jessie is now twelve years old in the seventh grade. Lamar is nine years old in the fourth grade. I am certain the children would love to stay with their favorite Grandmother."

Edward agreed and relayed Grandma Lily's request to his family. He was relieved that his mother could take care of the children while they went to California. He was ecstatic to travel with his wife although he would greatly miss his children. Edward and Felicity were leaving very soon in two days on Monday morning for California.

Within two days Jessie and Lamar moved into their grandmother's house carrying extra supplies and clothes. The children were frequently found at Grandma Lily's house. If they were not gardening, they would be playing their made-up fantasy games about living on other worlds. This fanciful play required much imagination with minimal physical activity making Lamar feel less helpless. He was recovering from leukemia and sometimes had little energy.

"Don't forget your medicine," said Felicity to her son Lamar. He was carrying another load of his clothes to Grandma Lily from his house.

"I didn't forget," replied Lamar. "It was in the first load and Grandma Lily already put it in the bathroom medicine cabinet."

Lamar liked to feel that he could take care of himself. Although he was thin and small, he carried his own supplies. His brown arms were strong and wiry his parents often said. He squinted his black, oval eyes that seemed to fill up his brown face giving him the look of a dreamer. A tousle of red-curly hair surrounded his face. He heaved the last load from his bed with a grunt.

His sister was standing at his bedroom door and saw her brother's effort. She ordered him, "Lamar, let me carry your things. The doctor said you have to take it easy."

Jessie was the practical type. Her full first name was Jessica. Everyone called her Jessie for a nickname. She was as tall as her grandmother, plumb and had curly, brown hair. She had inherited her grandmother's round dark-brown face, broad nose and full lips. Like her father and brother she had black, sparkling eyes that rarely held tears. She liked to wear blue jeans with plain t-shirts. She was seldom seen in a dress.

"Well, you can take half," replied Lamar relieved at his sister's offer.

Early Monday morning Felicity awoke to finish packing. She tied her straight, brown hair back in her usual ponytail. She was as tall as her husband, thin, with olive skin tone. The ponytail emphasized her wide-blue eyes, narrow nose and small mouth. Her clothes were out of date and she did not seem to mind that she was not at the height of fashion. Her happiness was with her family. She sang softly as she packed allowing Edward to sleep another hour.

The taxicab arrived in the afternoon to take the Johnsons to the airport. Grandma Lily from her front door watched her son and his wife load their suitcases in the yellow taxicab.

"Jessie and Lamar will be fine," Grandma Lily reassured the parents. "You just go and have a good time." The children stood quietly beside their grandmother already missing their parents.

Edward and Felicity departed for the airport joyfully waving to Grandma Lily and the children from their taxi.

It was dusk on the same day of the parents' departure when Felicity returned unexpectedly to Grandma Lily's house. She stepped slowly from the taxi without her husband. She walked alone through her house and through the back alley to Grandma Lily's garden. Felicity faltered when she reached the iron gate. Grandma Lily and Jessie had previously stopped working in the garden. They were engrossed in listening to Lamar reading poetry when Felicity surprisingly appeared and walked in through the garden gate.

Standing still at the garden gate Felicity in a very gentle voice said,

"Grandma…children…"

Everyone turned with looks of astonishment to see Felicity standing near the gate with her blue eyes shiny from tears.

Then Felicity spoke so softly it was like a whisper,

"I have very bad news. Children, Edward…your father…has passed away. Grandma your son is dead."

Grandma Lily gasped and dropped the trowel she was using to weed the garden.

"How…Why?" asked Grandma Lily.

"He passed away from a heart attack," replied Felicity reaching out to her children and hugging them trying to ease their sorrow.

"He said his chest ached and he couldn't catch his breath in the airport. An ambulance was called. They rushed him to the hospital. He died within an hour after being admitted."

After hearing this sad news, Grandma insisted that Felicity spend the night with her and also the children. Sadness marched through the house that night. Laughter became a stranger and was heard no longer at Grandma Lily's house or the Johnson's house next door after the death of the children's beloved father.

That night Felicity rejected Grandma Lily's offer to combine their money and live in Grandma Lily's house. Felicity's pride acted like a wedge. Her pride seemed to separate not unite the Johnsons and Grandma Lily.

"I'm trying to keep my family together," Felicity said when she refused Grandma Lily's offer. There seemed to be stress on the words "my family" as if Grandma Lily was not a part of them anymore.

Days passed and then weeks passed with many changes. Grandma Lily became grayer and seemed smaller. She seldom worked in her garden. Eventually, she stopped the once delightful pastime of puttering around in her garden. Her contentment in the beauty of her garden disappeared.

"I feel worthless," said Grandma Lily to a friend she met on a bus.

Grandma Lily further explained, "I can't help my daughter-in-law and my grandchildren because I am on a fixed income being a retired elementary school teacher."

"You can try to comfort them," said the friend, "that costs no money."

"How can I console my son's family when I can't even comfort myself," replied Grandma Lily. "This grief is like a binding cord constricting me."

The garden was neglected. It was the place where Grandma Lily last saw her son and it was where she heard of his passing. The brilliant-colored flowers became dried stiff-brown stalks. The bees no longer came to the garden. There was no humming, reading out loud, playing fantasy games or laughter in the garden. There was no garden this season

Felicity also experienced many changes. Felicity worked longer hours at her job as cashier in a supermarket to take care of her family's needs of food, clothing and other daily necessities. She had to pay the mounting medical bills for Lamar.

"Darlings, I don't have any money for that right now," said Felicity to her children asking for money on school projects. "Maybe next week when I'm paid."

Felicity's hard work was not bringing enough money into their house for all the families' necessities. Without Edward's income, the family began failing. He had no life insurance policy to help support his family after his passing.

Then, Mr. Stimp, appeared at Felicity's job. He was a tall muscular man whose skin tones alternated between greenish-white and purplish-white colors depending on how much he was yelling. His pale-yellow hair was almost white and he had sharp features. He had small, blue eyes, a sharp-pointed nose and a mouth that turned sharply downwards.

Mr. Stimp flung open the door to the market where Felicity worked and stomped across the floor to her cash register. He callously interrupted Felicity scanning a customer's groceries.

"Are you Felicity Johnson," Mr. Stimp shouted.

"Yes," replied Felicity.

"Well, you owe me a bundle," yelled Mr. Stimp thrusting a handful of papers under Felicity's chin.

Felicity choked, stopped scanning groceries and lowered her head in dismay at Mr. Stimp asserting that she owed him money. She recovered quickly and asked her supervisor for a short break. Felicity requested that Mr. Stimp move to a quieter location in the market.

In the employees' break room, Mr. Stimp yelled, "Your thief of a husband owes me ten thousand dollars."

"Edward is dead," replied Felicity.

"I know that, what do you think I am a fool?" shouted Mr. Stimp. "He still owes me ten thousand dollars. He never paid me one cent. He put up his house as collateral. See look at these papers."

Mr. Stimp shoved more papers at Felicity. She hesitantly took the papers carefully reading the official documents.

"See, see," yelled Mr. Stimp stabbing his finger at the documents. "You have three months to pay me the money or I take the house. It is that simple."

Mr. Stimp left the papers with Felicity and stomped out of the room.

Felicity was devastated. She went home to the children trying to pretend that everything was fine. However, there were long silences in her conversation with the children. She would hesitate when the children asked her for money to buy food. Felicity was even reluctant to see Grandma Lily only visiting her once a week with the children instead of daily as the Johnsons had done previously.

During the infrequent times the children now visited their grandmother over the next month, they entered the house through the blue front door knocking loudly to be heard. They did not go as they had many times before through the back wooden gate in their yard, along the short walkway protected from the street by a locked heavy-metal gate, through the garden's iron gate and walk directly into Grandma Lily's garden. This path was too painful to use. The walkway reminded the children of their now deceased father, so they stopped taking it to their grandma's house.

One day, two months later, near the end of summer and near the start of the school year, Lamar visited his grandmother with his sister. He had not been in the garden for several weeks and he had an urge to view it hoping for the previous better times. There were dull-grey drapes hanging over the picture window that led into Grandma Lily's garden. The somber-gray drapes fully covered the window. Lamar wondered what happened to the colorful orange and yellow- patterned drapes that had previously lined the picture window.

Lamar parted the grey drapes to view the garden. He expected to see a lonely place containing dried flowers and weeds. He was startled to see a large-black cat with an uneven cream-colored spot on its forehead curled up asleep in the middle of the garden surrounded by dried plants and bare patches of ground. The cat was as big as a terrier dog. It was motionless and Lamar thought it was a statue.

Lamar clapped his hands loudly to see if the cat would react to the sound. The boy watched intently as the cat moved its head to look directly at Lamar with its green eyes. Then, the cat winked his left eye at Lamar, raised his right paw and waved hello to the inquisitive boy.

"Grandma," shouted Lamar after viewing the cat's antics. "There's a cat in the garden and it waved hello to me."

Jessie rushed from the kitchen where she was making oatmeal cookies with Grandma Lily. She hurried to stand beside her brother at the picture window to see the cat. She also found that a cat in the yard was an exceptional sight.

"Where did this cat come from Grandma? What is his name?" asked Jessie bewildered at seeing a cat in this particular place.

"Oh, his name is Vesper," said Grandma Lily from the kitchen that was directly attached to the living room.

Grandma Lily continued preparing the cookie dough adding rolled oats to the dough mixture and explained, "One of our neighbors had to return to her home country in Europe. She'll be gone for several months. She needed someone to take care of her cat. I volunteered. I guess I'm just getting too lonely."

"Oh, the cat was in a circus and knows all kinds of tricks," Grandma Lily said further.

Felicity did come over to see the new cat when the children told her about Grandma's new animal resident. The children came over to see Vesper frequently at first. However, discouraged by the cat's inactivity, the dismal garden and Grandma Lily's dampened spirits, Felicity and the children visited Grandma Lily less and less. The novelty of Grandma having a cat wore away quickly.

The cat had a mind of his own refusing to perform anymore tricks. He stayed in the garden or in the house curled in a ball asleep. When it was time to eat, he awoke and took mincing steps to his bowl of food in the kitchen. After eating, he meticulously groomed himself with his pink tongue cleaning and smoothing his hair. Then, he returned to sleep.

A couple of weeks after Vesper arrived, the cat was awakened from his sleep in the living room by the loud sounds of a man yelling outside Grandma Lily's front door. The shouting echoed through the house jolting the cat from his slumber. Vesper discontented by the disagreeable noise stood on all four paws with his ears flicking back and forth. Who was making these loud noises?

The person was Mr. Stimp. He had decided to see the Johnson house to make certain the house was still in good repair. He saw Felicity, Jessie and Lamar standing outside Grandma Lily's house. Felicity's right hand was raised to knock on Grandma Lily's front door.

Mr. Stimp said loudly to Felicity, "You need to pay me within thirty days. You owe me ten thousand dollars"

Grandma looked out her front door window to see what the commotion was. She immediately opened the front door when she saw the Johnson family being badgered. Grandma Lily quickly beckoned Felicity and the children into her house.

"I need my money in thirty days," yelled Mr. Stimp as he followed the Johnsons inside Grandma Lily's house. He looked purplish-white.

"I still have time," responded Felicity.

"Ha, you don't have any money to pay me," shouted Mr. Stimp.

Felicity quieted Mr. Stimp by speaking firmly to him or maybe he became tired of hearing himself shout. Felicity finally revealed who he was to Grandma Lily and the children. She explained why he was demanding money from her.

Grandma Lily and the children were horrified at Felicity's revelation. Grandma Lily ordered Mr. Stimp to leave. He did not obey Grandma Lily.

Mr. Stimp instead moved to the picture window. The grey drapes were drawn displaying the garden with dried plants. Mr. Stimp pointed to the garden laughing derisively at the puny plants in the garden.

His close presence to the garden caused the plants to shrivel more with the bare patches in the garden becoming wider.

"There will be no garden this season," said Mr. Stimp. He was mocking Grandma Lily's desolate garden. He left the house vowing to return for his money or with an eviction notice for the Johnsons to vacate their house.

After Mr. Stimp departed, Grandma tried to reassure Felicity that everything would be fine.

"Felicity, don't worry. We will find a way to get the ten thousand dollars," said Grandma Lily reassuringly to her daughter-in-law sobbing.

"How, Grandma Lily? How can you help me?" Felicity responded tartly. "Do you have ten thousand dollars?"

Grandma Lily shook her head side-to-side indicating "no".

"Then you can't help me," Felicity said.

Grandma Lily watched helplessly as Felicity stifled her tears, gathered the children and left saying a perfunctory good-bye. Grandma Lily did not know how to make things right. Her solicitous attempt to thwart Mr. Stimp's threats was unsuccessful.

Autumn came bringing chilly winds and green leaves of trees changing colors to scarlet, gold and orange. Autumn changed into winter. Winter was more dismal. Felicity was able to get a stay of nine months on their eviction due to her husband's sudden death. However, the cold air outside crept into the lives of the Johnsons.

For Jessie and Lamar going to school was a chore. The happiness of going to school was drained away. Their father was not there to listen to how their day went in school, laugh at funny stories or help with homework.

The holiday season was not welcomed this year. The joy of giving was not important and the two-week holiday from school was dreaded by the children. They had made a secret pact with each other to pretend to be happy by not revealing their real feelings of unhappiness.

Grandma became despondent worrying about her daughter-in-law's financial predicament. She felt less energetic and slept poorly at night. She began taking daily naps at noon to renew her energy. Today she was especially tired and fell asleep quickly. She dreamed that her garden was in the midst of major magical events. Grandma Lily dreamed that when she left her house to go on errands and Vesper was alone, he would magically disappear from her house.

Vesper would reappear in his true world of Alarn, a world of magicians that existed in another dimension. Vesper came to Earth because he was a Guardian of Alarn magic. As a Guardian, he could talk.

"Again, Vesper," said the Head Guardian of Alarn. "There is another errant wizard misusing magic on planet Earth." The Head Guardian was leaning over a table where a large globe was pulsating light. Scenes from different worlds could be seen between the pulses of light. The Head Guardian's flowing, black robe partially blocked Vesper's view of the globe.

"Who is it this time?" asked Vesper his lithe-feline body wrapped in a ball on a dark-green sofa in the Head Guardian's untidy study. The room was filled with magical gadgets and bookshelves with volumes of books containing magic lore.

"Altere Stimp," responded the Head Guardian, "has somehow interfered with a magic place. Vesper, can you go to Earth now?"

Vesper shook his head in the affirmative. He jumped from the sofa and padded over to the Head Guardian to receive more information about his assignment.

The Head Guardian explained as he viewed the happenings in the globe, "Vesper as you know Earth has no natural magic within the makeup of the planet. You know how Alarn magicians like to practice magic on Earth. This time an Alarn magician did something good. The magician infiltrated a plot of ground with magic, beauty and joy. The garden cared by Grandma Lily Johnson is such a place."

"So what do I need to do?" asked Vesper unclear about helping a good practice of magic.

The Head Guardian responded, "Altere Stimp has siphoned away the magic from the garden to be used in some nefarious scheme. So the garden is badly harmed. If the magic, beauty and joy is not restored in the garden then adjacent properties will also lose whatever beauty and joy these areas have."

"Like a row of dominoes falling," said Vesper.

"Correct," answered the Head Guardian. "The despair and anguish from the dried stunted plants of Grandma Lily's garden can adversely infect other areas of surrounding land. The natural beauty and happiness of these properties would deteriorate and finally disappear."

The Head Guardian continued, "The loss of magic in Grandma Lily's garden will then spread throughout the world. The entire planet Earth will eventually lose its magic and happiness. The Earth world will become a place of misery and doom if evil Altere Stimp succeeds."

"Is my job to restore magic, beauty and joy in Grandma Lily's garden and save the planet Earth from doom?" asked Vesper.

"Yes," replied the Head Guardian.

Shortly after hearing the details of his assignment, Vesper appeared on Earth in the guise of an ordinary household cat. He was able to befriend an intuitive person who somehow felt that Vesper was special. Vesper maneuvered and found a way to enter into Grandma Lily's house as a temporary, guest pet. He still had a major problem because of Altere Stimp's action of removing the magic, beauty and joy had badly damaged the garden possibly beyond repair.

After investigating the situation, Vesper returned to Alarn to discuss the predicament of losing magic in Grandma Lily's garden. He spoke to renowned magicians of Alarn that practiced good healing magic. Some of the magicians suggested removing part of the energy from an existing repository of magic on Earth like Excalibur Sword or the Dome of Ries. This suggestion would need a large amount of energy. There were many other suggestions that would require particular spells or incantations. No magician was one hundred percent certain their suggestion would work.

Vesper decided to try the simplest suggestion made by a magician knowledgeable in mountain magic, Onel. The magician was digging a cave in the side of a mountain to help the bears in hibernating when Vesper asked his advice.

"Heelllooo," said Vesper's echo when he spoke down a shaft trying to find the magician.

"Hello, back to you," said Onel. He suddenly appeared at the mouth of the shaft with his black robe flapping in the breeze and his pointed hat askew on his head.

"I need to get a restoring spell to return magic, beauty and joy to a garden," said Vesper.

"Onel replied, "Hmmm, what about having a magical being walking through the garden leaving prints. During the next rainstorm, the prints would fill with rainwater and by walking around the area three times thinking positive thoughts then magic, beauty and joy would return to the garden."

Vesper thanked Onel and looked for his friends that were magical beings. Vesper knew magical beings such as Garlen, the blue centaur, Rus, the golden phoenix, and Arn, the silver unicorn. When Vesper asked his magical friends to help with the garden, they apologized profusely. They were already engaged in vital magic projects and could not assist him.

Although Vesper dealt with magic, he was not considered a magical being like a centaur or a unicorn. At one time Vesper was a lifeless statue. An Alarn magician's powerful gold medallion was attached to Vesper. The magic medallion gave Vesper the ability to think, move and behave like a real cat. The magic medallion also gave Vesper the cat the power of human speech. The uneven cream-colored mark on Vesper's forehead was a remnant of the medallion.

It was with these serious considerations that Vesper decided he would have to reveal his true purpose to Grandma Lily and the Johnsons. On the first day of the children's winter vacation from school, Grandma Lily and the Johnsons had dinner together in Grandma Lily's house. After dinner, everyone including Vesper congregated in the living room where the picture window was that led to the garden. The dull-grey drapes covered the window blocking a view of the garden.

Vesper walked to the grey drapes and pulled the drawstrings with his paws revealing the garden. It was nighttime. The garden was covered in snow with a full moon glowing in the dark sky. No remnants of plants could be seen. Grandma Lily had placed solar-powered garden lights around the circumference of the garden. They looked like miniature, colonial street lamps. The yellow glow from the garden lights and the silver moonlight from the full moon showed the hills and valleys of the snowdrifts in the yard.

Vesper cleared his throat and began talking.

"Grandma Lily, Felicity, Jessie and Lamar," Vesper said turning to face each person as he said their name. "We are faced with a dire dilemma."

"I knew he could talk," said Lamar excitedly. The others were not as jubilant. They were surprised and dismayed.

"Is this a circus act?" asked Felicity of Grandma Lily.

"No, it is not a circus act," replied Grandma. Somehow Grandma Lily had suspected that Vesper was unique and different from her first meeting with this unusual cat. Therefore, she was not quite as surprised at Vesper talking. There were times when Grandma Lily found Vesper staring at her as if he were trying to tell her something. His mouth would move like he was mumbling to himself and she thought she heard him speak words. At these times, it was easier for Grandma Lily to believe the cat was doing circus tricks and not actually talking.

Felicity was shocked that a cat could talk. She had read fairy tales as a child but never believed they could be true. She sat through the ensuing discussion making few comments.

"What is the dilemma?" asked Jessie. She was like her grandmother and was initially suspicious of Vesper. He was just too wise. Once she was crying by herself in the living room of Grandma Lily's house saddened by the death of her father. Vesper crept near her and his purring calmed her as Jessie petted the cat. After that incident, whenever Jessie became despondent about her father, she pretended a black cat with an irregular cream-colored mark on his forehead was near her purring. Her melancholy would turn to understanding that people and things cannot last forever.

Vesper talked about his real identity of being a Guardian of Alarn magic. He thoroughly discussed the garden's predicament of not having any more magic, beauty and joy. He had to find a solution for the loss of these things.

"Why must the garden have magic in it?" asked Jessie.

"It's like weaving a piece of cloth," responded Vesper. "The cloth can be woven with certain strands of color like red, blue and yellow."

"If another color is added like green," continued Vesper. "Then you have a pattern of red, blue, yellow and green."

"I know," interjected Lamar. "If you take away one of the colors like green you have a different pattern."

Jessie thought like the Johnson family were woven together. When her father died, an important strand was gone and the family unraveled.

Felicity was doing her best to follow the conversation. She felt like she was on a steep roller coaster ride in the amusement park that was speeding very fast up hills, down hills and around curves. Felicity feeling faint had to lie down on the sofa.

There were so many new concepts to understand. First, that magic even existed. Magic was just a made-up fairy tale. Second, that an animal could talk. Maybe whatever Vesper was inside should not be confused with how he appeared on the outside.

Felicity gained control over her senses. She sat up and asked, "So what is magic?"

"And is magic scientific?" asked Grandma Lily.

"On Alarn we see magic as using thoughts, words and actions to change something," replied Vesper. "Magic is neither good nor bad. It is like electricity or magnetism, another type of energy."

Lamar then inquired, "I wonder if magic can help us save our house from Mr. Stimp?"

Felicity was too embarrassed to inquire out loud what her son had asked. Magic was just pretend as any adult knows. Yet Felicity could only hope that bringing the garden back to life would also bring happiness back into her family's life.

"I neglected the garden because it reminded me of my son," confessed Grandma Lily. "I don't think it was due to the loss of magic. So if I begin to care for the garden again the plants will flourish. Their beauty and joy will return."

"I think all of us need to care for the garden," said Felicity. Everyone agreed and plans were made on the design of the garden especially what flowers to plant in the spring.

Jessie was the first to notice outside in the garden, the gold and silver beams of light were dancing across the mounds of snow. The beams merged together to form the silver shape of a horse with a gold, spiral-shaped horn on its forehead and silver hooves.

"Look!" exclaimed Jessie pointing to the garden and directing everyone's attention to the remarkable phenomenon of a unicorn appearing in the garden.

"Arn," said Vesper. "The unicorn came to help us. She must have finished the magic job she was working on earlier than expected."

The unicorn trotted to the picture window and looked inside at everyone assembled in the living room. She tossed her head as if to say yes making her silver mane circle her head like a crown from this movement.

"Oh, can she come inside?" asked Jessie of Vesper. She was enraptured by the beauty of the magical animal like the rest of the room.

Grandma Lily was relieved when Vesper explained that unicorns usually prefer to be outside. They did not like the confines of an enclosed building. Grandma did not want a horse or a unicorn tracking up her clean carpet with wet hooves inside her home.

Vesper was happy that Arn would help bring magic, beauty and joy to the garden again. Vesper nodded goodbye to her as Arn turned away from the delighted group of people. She left the garden through the iron gate that swung open by a touch of Arn's horn.

Winter past and spring came with the Johnsons able to remain in their house past the nine month deadline. There were some technical mix-ups with Mr. Stimp's documents in the courts giving the Johnsons additional time to stay on their property. They received twelve more months to find funds to pay Mr. Stimp.

Arn, the silver unicorn, did help Vesper. She reappeared in the garden on a warm, spring night during a full moon trotting back and forth across the ground several times. Each hoof print left silver prints that glowed in the darkness of the night. Only Vesper and Grandma Lily witnessed this magnificent spectacle.

The next day it rained filling the hoof prints with water. After the rainstorm ended, Vesper and Grandma Lily walked around the garden three times thinking positive thoughts of happiness and beauty. Each time they circled the garden, the silver water in the hoof prints shone brighter.

With the ground still wet from rain, Felicity, Jessie and Grandma Lily carefully planted flower shoots in the damp soil. First, they dug a deep hole in the brown soil with a trowel. Gently they placed the delicate shoots of flowers purchased from the local plant vendor in the ground firmly covering the roots with damp earth. The plants were sprayed with water from a green-worn garden hose and given plant food.

Along with the flower shoots, Jessie planted her wish that her family's problems be solved. The children decided that it would do no harm and possibly good if a wish was planted in a magical garden. Jessie dug a hole with a trowel. She pretended she was placing something beautiful and joyful in the hole and covered the hole with the brown, damp earth.

Vesper went inside the house to watch the garden being planted from behind the picture window. The cat saw Jessie's actions and said nothing. Vesper using his exceptional hearing, listening with the rest of the gardeners to Lamar reading poetry from the green patio chair. The magical cat was impressed by the hard work of Grandma Lily and the Johnsons in planting the garden.

The gold, purple, pink and silver flowers grew rapidly in Grandma Lily's garden. The bees returned and laughter sounds could be heard intermixed with the humming of the bees. Also, magic returned to Grandma Lily's garden. Vesper's job was done. It was time for Vesper to leave.

The goodbyes between the Johnsons and Vesper were both happy and sad. They were happy because the Johnsons were woven together again as a family and no longer unraveling. A unicorn had trotted through the garden. Magic, beauty and joy were again in the garden. The sadness was in Vesper leaving the Johnsons. Vesper was moving to another city with his benevolent owner to handle more issues concerning magic.

Before the owner returned, the Johnsons gave Vesper a farewell party in the garden one summer evening. Arn came to the party in her silver magnificence to walk around the garden one last time. Lamar wrote a poem for the party expressing his feelings and read it to the appreciation of everyone.

Felicity was late for the party and was able to hear the end of her son's poem on the virtues of family and friends uniting for magic. Felicity explained why she was late.

"I was searching through papers in Edward's messy desk to find the house deed," said Felicity.

She held up a few wrinkled papers and excitedly said, "I found a fifty thousand dollar life insurance policy on Edward. I've searched that desk many times and never saw this policy. Now we can pay Mr. Stimp the ten thousand dollars owed. We can keep our house." Everyone laughed and hugged each other at the good news.

Was the discovery of the life insurance policy due to magic because Jessie had planted a wish in Grandma Lily's magic garden? Maybe it was not magic. The policy could have been intermingled in the many papers covering Edward's desk. Vesper knew the truth, however he never told it. Sometimes, believing in magic, beauty and joy is just as important as there actually being magic.

At the end of the party, Grandma Lily asked Vesper, "Please visit us again and see how the garden is doing."

"Grandma Lily, I shall try to return for a visit," replied Vesper

"I also have another wish. I wish there will be a garden every season," Grandma Lily hoped. With the help of Felicity, Jessie and Lamar there was a magical garden filled with beauty, love and joy planted each year.

However, Grandma soon discovered this Vesper story all took place in Grandma Lily's dream. Finally, Grandma Lily awoke from her strange dream about the cat, Vesper, to find Felicity standing beside the sofa where Grandma Lily was taking a nap.

"I must have really been tired, I have been asleep for hours," Grandma Lily said to Felicity while gazing at the clock on the wall that now read five o'clock.

Felicity excitedly said, "Grandma Lily I have something important to tell you. I just discovered a fifty thousand dollar life insurance policy on Edward that was in his desk. We will not lose the house after all!"

"Felicity this is wonderful news! However, I just had a strange dream that through magic, love and joy we were able to save your house," Grandma Lily happily responded.

Grandma Lily decided at that moment not to worry about the past. She would enjoy life again. She decided to have a special garden this season and every season thereafter with the help of her wonderful daughter-in-law, Felicity, and kind, loving grandchildren, Jessie and Lamar.

"There will certainly be a wonderful garden this season!" Grandma Lily said each year to her neighbors and friends.

Each planting season thereafter the bees could be heard humming and seen flying between the colorful flower beds in Grandma Lily's garden. Grandma kept her promise and made sure there were beautiful, magical flowers planted in her back yard each year. Never again did Grandma Lily ever utter the words, "No garden this season," because joy and happiness resounded in her garden and inside both Johnson homes regularly.

About the Author

Hazel Ann Lee is a contemporary writer of poems, novels and short stories relating to nature, humanity and science. She is a scientist and teacher. Since H.A. Lee was a child, she has found science

to be an interesting and exciting subject to study. Learning about the lives and careers of many admirable scientists and reading many books about science, nature and science-fiction led H.A. Lee to becoming accomplished as a writer. She also enjoys gardening. and diligently cares for her quaint-city garden year-round. In addition, she especially likes to care for her astute orange tabby cat.

H. A. Lee's poetry has been published by the Moonstone Arts Center in their *21st and 22nd Anthology Editions of the Poetry Ink Collection.* She is furthermore an award winning lyricist and songwriter. Her talents include writing the librettos for two contemporary operas, *Partway To Freedom* and *A Sailor's Civil War Tale: Let Courage Be The Light.* In addition, H.A. Lee has two books about science fiction and fantasy soon to be published titled *Vesper* and *Strivers.*

Printed in the United States
By Bookmasters